Felipe Klein Fiorentin

Numerical simulation of the behaviour of thin parts during milling

AF550930

Felipe Klein Fiorentin

Numerical simulation of the behaviour of thin parts during milling

Dynamics and vibrations in machining

ScienciaScripts

Imprint

Any brand names and product names mentioned in this book are subject to trademark, brand or patent protection and are trademarks or registered trademarks of their respective holders. The use of brand names, product names, common names, trade names, product descriptions etc. even without a particular marking in this work is in no way to be construed to mean that such names may be regarded as unrestricted in respect of trademark and brand protection legislation and could thus be used by anyone.

Cover image: www.ingimage.com

This book is a translation from the original published under ISBN 978-620-6-75640-8.

Publisher:
Sciencia Scripts
is a trademark of
Dodo Books Indian Ocean Ltd. and OmniScriptum S.R.L publishing group

120 High Road, East Finchley, London, N2 9ED, United Kingdom
Str. Armeneasca 28/1, office 1, Chisinau MD-2012, Republic of Moldova, Europe
Printed at: see last page
ISBN: 978-620-7-68975-0

Copyright © Felipe Klein Fiorentin
Copyright © 2024 Dodo Books Indian Ocean Ltd. and OmniScriptum S.R.L publishing group

Contents

ACKNOWLEDGEMENTS

I would like to thank my mum, Maria Assunta Klein Fiorentin, for her unconditional support and for my professional qualification.

To Prof Dr Thiago Antonio Fiorentin, who closely followed and guided all the stages of this work, from the theoretical development to the experimental procedure, adding profoundly to the work with suggestions and additional information.

Finally, to the Federal University of Santa Catarina and the National Research and Development Council, for the opportunities to apply the knowledge gained at university through scientific initiation.

SUMMARY

In naval engineering it is common to apply slender handles, such as propeller blades and heat exchanger fins. These handles often undergo a machining process, either roughing or surface finishing. Because these components are relatively flexible, machining them is a delicate process. In practical applications related to machining a handle, the resonance frequency is always an important parameter and must be taken into account. Material removal is inherent to any machining process, so there are variations in the mass of the handle. In addition, as the geometry of the handles changes, variations in rigidity are also present. These two parameters determine the natural frequency of an object and, together with the damping factor, also determine the resonance frequency. Another factor that influences the rigidity of the assembly is the fastening system. You should avoid exciting these components at frequencies close to the resonance frequency. In order to minimise grip vibrations in the machining process and consequently improve the surface finish, this study aims to use numerical simulation using the finite element method to predict the dynamic behaviour of a given system by analysing its vibrations. Numerical simulations were carried out with proprietary codes, both for the time domain and the frequency domain, using a beam element. By correlating vibration amplitudes with roughness, it was possible to estimate the surface finish of the machined grip.

1 INTRODUCTION

Manufacturing processes are the economic basis of an industrialised nation, and in general the degree of economic development of a country can be determined by its level of industrialisation and innovation. In recent years, new technologies have been developed and added to manufacturing processes, mainly systems linked to process control and automation. The growing demand for industrialised products makes it obligatory for any industry to continually improve its processes, reducing manufacturing time and the number of defective handles.

Among the various manufacturing processes used in shipbuilding and engineering, machining is a very important process, especially when it comes to finishing surfaces. When this finishing takes place on very flexible handles, small forces can cause large displacements, and minimising these displacements is vital in the manufacturing process, otherwise the desired surface finish will hardly be obtained.

Among the relatively flexible surfaces that undergo a finishing process during machining are those of some propellers, which are aggravated by their complex geometries (JAE-WOONG YOUN, 2003). The surface quality obtained by the milling process is directly linked to the efficiency of the propellant (KUO and DZAN, 2002). Parts of heat exchangers, such as plates or fins with special geometries are also often subjected to the milling process, and although they do not have such strict dimensional requirements as propeller blades, they are extremely flexible and difficult to machine (WIPPLINGER, 2006).

The machining process has received a great deal of attention in recent years and although it is a procedure known for its high cost, it is often necessary (GRZESIK, 2008). Handles that are cast, shaped or obtained by other processes use machining to obtain a more refined surface finish, so this process is present in a huge number of products, even if it is only used as a finishing process.

There is a current market demand for components with high geometric, dimensional and surface quality tolerances, which would have seemed unattainable just a few years ago. Meeting this demand is made possible by developments in manufacturing processes. Components whose manufacturing costs were prohibitive until recently are now mass-produced at relatively low costs, largely due to advances in manufacturing control methods (STOETERAUS, 1999). In this context of tighter tolerances, machining is the dominant process for obtaining handles with a high surface finish.

It is becoming increasingly common to try to predict the outcome of a given process, eliminating the need to physically carry out the procedure to find out whether the result will be satisfactory or not. The biggest incentive for these endeavours is undoubtedly the financial aspect: it is extremely economically unviable to carry out a certain procedure and then find that the resulting handle has some parameter outside the specification. For simpler cases, analytical calculations are able to predict some behaviour. However, in general industry, problems of a relatively complex geometry and nature make analytical procedures unfeasible, and numerical simulation is used as a solution.

Simulation and modelling are strong trends in the areas of scientific, technological and industrial research, due to the advantages they can offer in terms of reducing costs and the time it takes to complete a project.

Simulating a machining process consists of representing and analysing it numerically with the help of a mathematical model, which should ideally be as close as possible to the real thing (DOMINGOS, 2002). Obviously, making a simulation increasingly assertive often involves increasing simulation times, since the number of equations and phenomena involved is greater (PIMENTEL, 2011). It is up to the engineer to define the necessary simplifications, quantifying which phenomena are really significant in a simulation.

However, numerical simulation, like any other scientific method, needs to have its results validated. In the case of machining, the experimental procedure is the way to do this. This experiment must contain the same parameters as the simulation, in order to make both equivalent and consequently with similar results. The most common parameters to be taken into account in machining are the rotation of the handle or tool, depth of cut, machining direction and rake.

As an experimental procedure was needed to validate the analytical and numerical simulation results, the turning or milling process was considered. We opted for the milling machining operation, since it would be possible to carry out such an experiment.

Some fundamental pillars will be used as theoretical references: the Euller-Bernoulli beam model, an adaptation of Altintas' shear force model and Hubolt's time integration method.

1.1 Objectives

The aim of this work is to develop a finite element computer code capable of predicting the dynamic behaviour of a given system during the milling process. This model will be validated with the help of other simpler numerical models, such as modal analysis. A definitive validation could be carried out through experimental procedures; parameters such as vibration amplitudes during the manufacturing process and the roughness of the manufactured handle could be studied. The surface finish of the machined handles could be compared with that predicted by the numerical model

1.1.1 General Objective

Developing a model capable of predicting the surface finish and dynamic behaviour of a given system during a milling process.

1.1.2 Spert Objective

- Implement a model that is able to reproduce the cutting force over time.
- Develop a computer code capable of representing the machining phenomenon, representing the variations in handle geometry over time.
- Validate the suggested numerical model.

1.2 State of the Art and Background

Predicting the dynamic characteristics during the milling process is essential. For the manufacture of a given component to be effective, the stability of the operation must be guaranteed. A study was carried out on the stability of a machining operation in the time domain (ZHONGQUN and QIANG, 2008). A very interesting approach is presented, using the instantaneous chip thickness to build a correlation with the effects of regenerative forcing. The authors construct stability curves equivalent to what the lobe diagram represents in the frequency domain.

As already mentioned, one of the most critical processes in precision machining is the machining of naval propeller blades. The best tool path in a machining process for these components was studied (JAE-WOONG YOUN, 2003).

One of the most vital steps in predicting the dynamic behaviour of the system over time is an adequate cutting force model. By observing the engagement of the tool with the handle throughout a rotation, Altintas (2011) built a force model derived from the chip thickness during the passage of each tooth. Since chip thickness is directly linked to the phenomenon of regenerative vibration, this method makes it possible to predict the stability of a given process over time.

Despite having very different specific purposes to this work, some studies use very similar approaches. One study using the Timoshenko beam model analyses the stability of a wind turbine (HANSEN, 2004). Another study, still using beam elements, was developed to analyse the dynamic behaviour of a given system (M. JURECZKO, 2005). In this study, the time-varying force comes from the wind, but the methodology applied is similar to this work, despite the fact that they deal with very different subjects.

This work is part of this context by proposing a prediction for a flexible system with excessive vibrations, using a numerical model to determine the regions with the worst surface finish, as well as predicting the system's dynamic behaviour for certain cutting conditions.

2 THEORETICAL BASIS

2.1 Vibration

Vibration is an inherent feature of any process that has oscillations in the forces involved over time, and the milling process has this characteristic. Because they are factors arising from the nature of the process, vibrations cannot be eliminated, only attenuated. In finishing processes, or even roughing processes that do not have other processes in sequence, there is great concern about surface quality and dimensional parameters, as a result of which vibration movements are one of the critical variables (SCHUKZ, WURZ and BOHNER, 2001).

Vibrations in the grip, tool or both tend to generate rougher grips. Figure 1 illustrates the profile of a machined surface without vibrations. Figure 2 shows the profile for a vibrating tool. Although stability in a machining process will be discussed in the course of this work, an initial concept can be drawn up. In machining, a stable process is one that produces a good surface finish and little tool wear, whereas an unstable process, due to high vibration amplitudes, is characterised by poor surface finish and much higher levels of tool wear (WERNER, 1992).

Figure 1 - Tool path without vibrations in the process

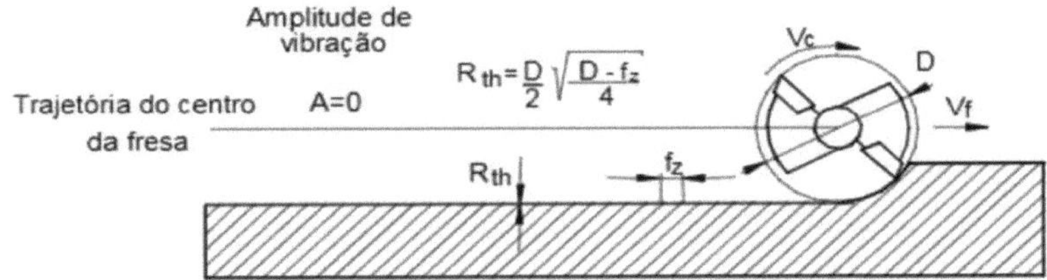

Source: Adapted from (WERNER, 1992).

Figure 2 - Tool path with vibrations in the process

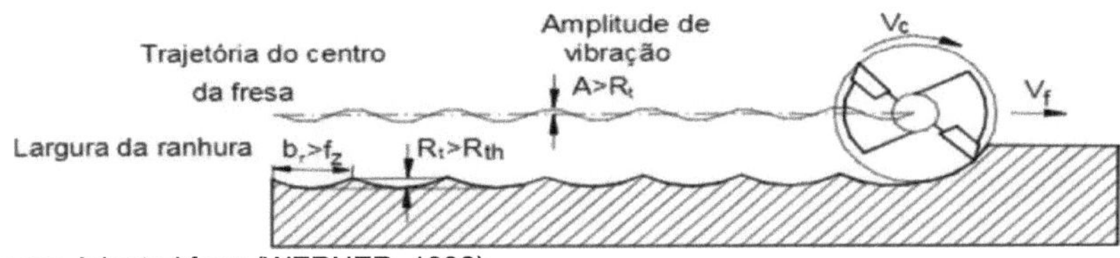

Source: Adapted from (WERNER, 1992).

2.1.1 Free vibration

Free vibration is the phenomenon of oscillation of a body which is not subject to time-varying external loads. The equation of motion for a system with one degree of freedom in this condition is

$$m\ddot{x} + c\dot{x} + kx = 0 \tag{1}$$

The equation above represents a system under damped vibration. A particular case of this system is for zero damping, where for an initial displacement,

$$x(t = 0) = x_0 \tag{2}$$

and for an initial speed,

$$\dot{x}(t = 0) = v_0 \tag{3}$$

you get the solution

$$x(t) = x_0 \cos(\omega_n t) + \frac{v_0}{\omega_n} sen(\omega_n t) \tag{4}$$

The movement represented in the equation above is a simple harmonic movement, it can be seen that the system will oscillate at a frequency ω_n, this is known as the system's natural angular frequency. A body with a degree of freedom that vibrates freely will always oscillate at a frequency equal to its natural frequency.

2.1.2 Forced Vibration

Forced vibration is defined as a mechanical system subjected to external loads, whether periodic or not. These loads usually come from unbalanced masses and the rotation of shafts and gears, for example (CREDE and HARRIS, 1961). A system with one degree of freedom subject to external forces can be modelled, as shown in Figure 3. The movement of this system is governed by the differential equation

$$m\ddot{x} + c\dot{x} + kx = f(t) \tag{5}$$

where *m is* the mass of the body, *k is* the stiffness of the spring and *c* is the damping, *f(t)* is the external force applied to this degree of freedom, *t is the* time and *x* is the position of this body.

Figure 3 - System with one degree of freedom and an external force

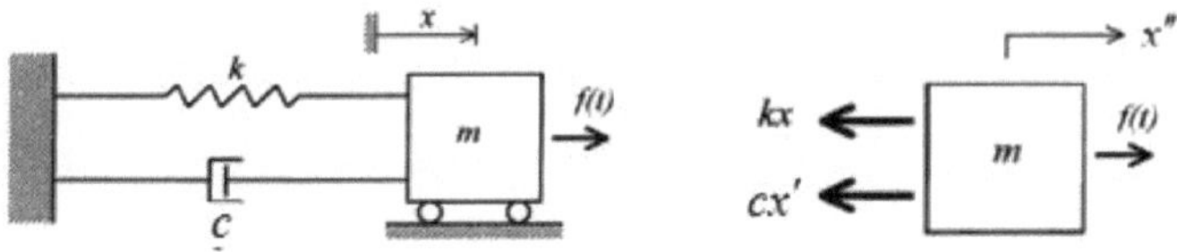

Source: Adapted from (EWINS, 1984)

A clever case would be for a harmonic force, in which the external force is represented as

$$f(t) = f_0 e^{i\omega t} e^{i\alpha} \quad (6)$$

where *a is* the phase, which is given by the angular position in a complex plane (ALTINTAS, 2000), ω is the excitation frequency and f_0 is the amplitude of this force. For such a harmonic excitation, the system response is given by

$$x(t) = X e^{i(\omega t + \phi)} \quad (7)$$

X is defined as the amplitude of vibration and ϕ *as* the phase, both are respectively

$$X = \frac{f_0}{k} \frac{1}{\sqrt{(1 - r^2)^2 + (2\varsigma r)^2}} \quad (8)$$

$$\phi = \alpha + \tan^{-1} \frac{-2\varsigma r}{1 - r^2} \quad (9)$$

where ς is the damping ratio and *r* is defined by the relationship between the frequency of the external force and the natural frequency of the system.

$$r = \frac{\omega}{\omega_n} \quad (10)$$

Forced vibration is inherent to the milling process, as it comes from the removal of material, however, in some cases this vibration is accompanied by another, known as regenerative vibration.

2.1.3 Regenerative Vibration

Regenerative vibrations, unlike forced vibrations, do not originate from external forces, but from a self-excitation mechanism, usually from periodic variations in chip thickness, which causes variations in cutting forces (ALTINTAS, 2000). The phenomenon occurs when a tooth passes through, generating a wavy surface, on which the next tooth passes, which in turn generates another wavy surface. Depending on the phase of these surfaces, an unstable process can occur. Figure 4 shows this phenomenon. The image shows two waves with a zero phase angle. As a result, the chip has a constant thickness, resulting in a constant cutting force.

Figure 4 - Chip formation with two waves in phase

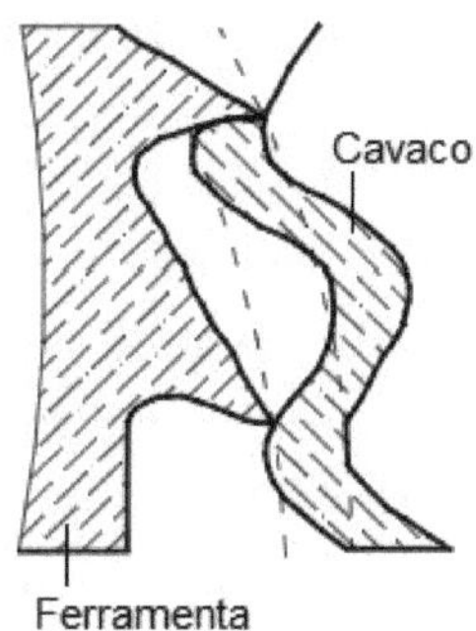

Source: (TLUSTY, 2000)

Figure 5 shows two waves offset by 180°, which is the worst possible situation, since sudden and periodic variations in chip thickness occur. When the peak of one wave coincides with the valley of the next, you have the smallest chip thickness, and in the opposite situation you have the maximum thickness. Such periodic variations in chip thickness result in large force variations (TLUSTY, 2000).

Figure 5 - Chip formation with two waves offset by 180°

Cavaco

Tool

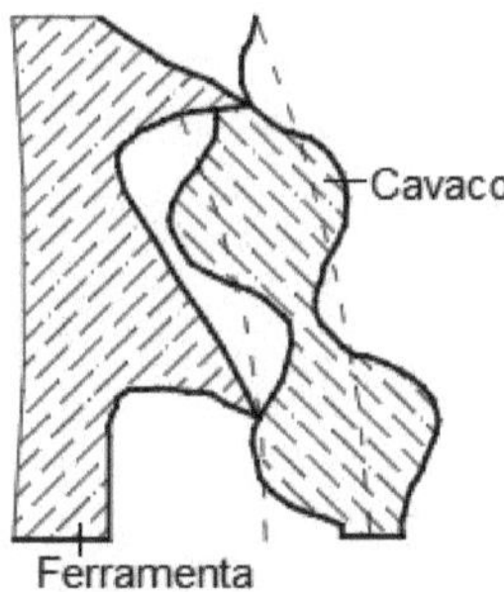

Source: (TLUSTY, 2000)

In a stable cutting situation, regenerative vibration is not allowed, so suitable cutting conditions must be defined, as well as the system's clamping conditions, in order to avoid such vibration. This stability of the process will be discussed in the course of this work, as well as the parameters involved and the solutions for minimising this phenomenon.

2.2 Milling

Milling is a specific type of machining process in which the tool is rotated and the material is removed intermittently (DROZDA and WICK, 1983). It is a method widely used to produce flat surfaces, contours, grooves and threads (DINIZ, MARCONDES and COPPINI, 1999). Milling has proved to be an extremely interesting process in the industrial sector, mainly due to the high material removal rates (KONIG and KLOCKE, 1999). Another essential feature of this process is its surface quality, even on handles with complex geometries (STEMMER, 1995).

Since the removal of material is intermittent in practically every milling process, each insert or tooth is either in contact with the handle or not, and as a result the thickness of the chip varies constantly (FERRARESI, 1977). When the tooth is not in contact with the grip, it does not produce any kind of force on it, and when it is engaged, stresses are applied to the grip, these oscillations in the presence of the force generate periodic loads along each rotation.

A characteristic that must be evaluated throughout the milling process is whether it is concordant, discordant or combined, Figure 6 illustrates these situations. Discordant milling is characterised by the rake being in the opposite direction to the cutting movements, which results in a chip with a theoretically zero initial thickness (KRATOCHVIL, 2004). In concordant milling, the cutting movements have the same direction as the rake and the cut starts with a maximum chip thickness. The combined process occurs when both situations are present, but only when the radial depth of cut, a_e, is greater than the radius of the tool.

Figure 6 - Discordant, concordant and combined or mixed milling

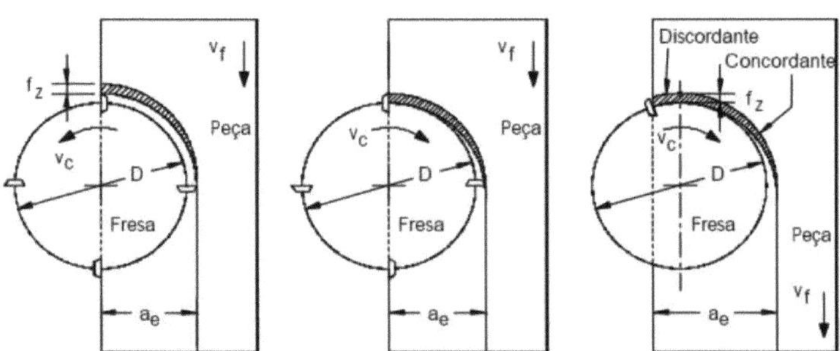

Source: (NCG, 2000)

2.2.1 Tangential cylindrical milling

The main characteristic of this process is a cutting plane parallel to the tool axis, and it is represented in Figure 7. This figure shows the concordant cutting process. This process is normally used to obtain flat surfaces (ALVES, 2016).

Figure 7 - Tangential cylindrical milling

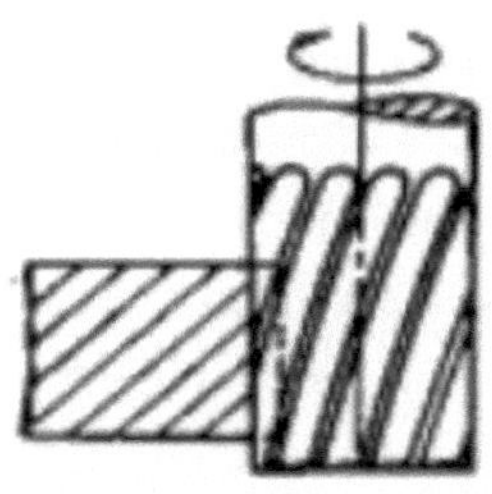

Source: (ALVES, 2016)

The decomposition of the machining force into radial, tangential and axial components will be shown in the course of this work, as well as its variations over time. After this decomposition, a change will be made to the global coordinate system for the x, y and z axes. This last step is necessary since the same axes and orientations must be used for the numerical model as for the experimental procedure.

2.2.2 Stability in the Machining Process

In order to meet the dimensional and surface finish requirements of a handle, the manufacturing process must be stable. An efficient process is one that is capable of producing a satisfactory number of handles within these tolerance limits, and one of the conditions that must be met is process stability. Various active and passive methods have been used to suppress excessive vibration.

Among the passive methods used to reduce or eliminate regenerative vibration is cutting speed oscillation (XIAO, KARUBE and SATO, 2002). This method is based on the idea that two waves with different periods are not out of phase. Some dynamic absorbers and vibration dampers are also widely used, techniques where several of these dampers are installed on parts of the machine tool (TOBIAS, 1965). Some studies on tool clamping have also been carried out; clamping devices involving rubber and rubber plates have been used to increase the damping of the system (RIVIN and KANG, 1989). However, these methods require knowledge of the system's dynamic response, since for certain cutting situations they can worsen the vibrations present.

Active methods are often used, for example studies employing piezoelectric actuators to suppress vibrations in the turning process (TARNG, KAO and LEE E.C., 2000). Some vibration absorbers actively acting on the cutting tool have been employed (LEE, NIAN and TARG Y., 2001), significantly reducing the vibration motions, accelerometers positioned on the tool provided information on the frequencies and amplitudes of the vibrations involved.

However, a step prior to treating regenerative vibration is detecting it. Ideally, one should be able to predict for a given system and given cutting conditions whether a system will be unstable or not. One of the most useful and widely used tools is the stability lobe diagram, which makes it possible to detect for a given system whether or not there will be stability at a given depth of cut and rotation (SCHMITZ and DONALSON, 2000). However, this method is limited to predicting the stability condition only in the frequency domain, which is not applicable in the study condition of this work.

Alternative solutions to the conventional lobe diagram have been created, for example, based on the measurement of the force signal, an algorithm automatically selects a suitable rotation to avoid regenerative vibration (SMITH and TLUSTY, 1990). Solutions using the time-domain integration method have been proposed (ZATARAIN, BEDIAGA, *et al.,* 2008), but when this method is compared with frequency-domain solutions it is extremely computationally expensive, requires much longer processing time and an automatic interpretation of the results is difficult to implement. A method for predicting stability in the time domain, taking into account the instantaneous thickness of the chip, in a situation where the tool is vibrating was studied (ZHONGQUN and QIANG, 2008), and comparisons with analytical solutions of simple models showed that the method used was efficient.

Although time-domain instability detection solutions are much more complex, they are often the only alternative. This work aims to implement a routine linked to time-domain simulation that is capable of predicting whether the process is unstable or not for a given situation and cut-off parameters, and for which situations self-excited vibration would be present.

2.2.3 Fastening System

The proper design of a clamping system is vital to the success of any machining operation. An efficient machining system is one that can hold the handle in a fixed position, but that the clamping forces are not too high as to cause plastic deformations in the handle (BOYLE, RONG and BROWN, 2011). Ideally, the clamping system should also be able to keep the handle in the same reference position, in which case reproducibility is important, as it is hoped that for a serialised process, clamping several handles will result in them all being in the same position (ASANTE, 2008).

In addition, in many cases it is desirable to minimise handle vibrations, and since the clamping system is directly involved in the dynamic characteristics of the system, a suitable design can significantly improve the quality of the process (ASADA, 1985). As mentioned above, knowing the precise position of the handle once it has been clamped is extremely important. One of the most common ways of minimising these positioning errors is to minimise the contact area between the positioners or supports and the handle, and this solution usually involves using a support system with spherical tips (JOHNSON, 1985).

Excessive clamping forces, in addition to generating deformations in the handle, also increase the contact area between the handle and the support, which is detrimental to the correct positioning of the handle (BAKERJIAN, 1992). Figure 8 shows a model of the contact area for a given force, where the grip surface is rigid and fixed.

Figure 8 - Contact area between the supporter and the handle

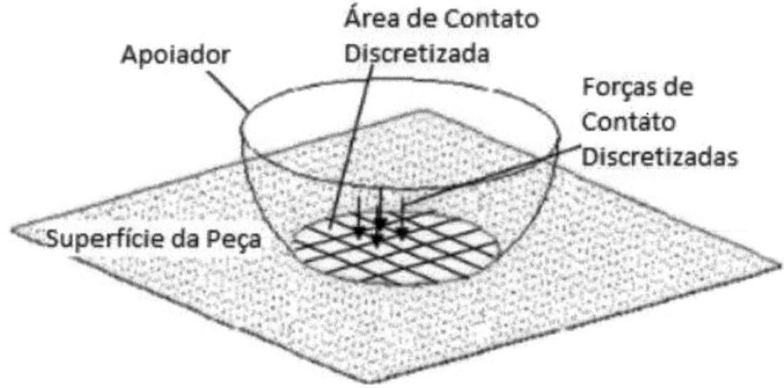

Supporter

Discretised Contact Area

Discretised Contact Forgings

Source: (ASANTE, 2008)

2.3 Roughness

Roughness is the parameter most often used to quantify whether or not a particular part has a good surface finish. It is often a design parameter, where a maximum allowable roughness value is specified for a given component. It is so widely used because it allows us to measure the quality of a particular surface finishing process. The measurement of roughness and the different parameters used to quantify it are related to the application of the project, such as the desired coefficient of friction, wear or adhesion. The latter, for example, is an exception to the others, since for good adhesion of paints and films, high roughness is desirable. There is no ideal or most suitable way of measuring roughness for all needs, which is why there are different ways of measuring roughness, one of the most widespread being Average Roughness, or *Ra*.

This method is widely used because it is easy to use and requires relatively simple electronic circuits to calculate. Average roughness is defined as the average deviation of a profile from its mean line. The expression that represents this roughness is given by

$$R^a = \frac{1}{l_m}\int_0^{l_m} |y(x)| dx \qquad (11)$$

where lm is the length of the measurement, and y is the amplitude, which is a function of position.

3 MATERIALS AND METHODS

3.1 Frequency Domain Simulation

Frequency domain simulations are widely used to represent the machining process, mainly because the excitation forces are usually periodic, and can be easily represented in the frequency domain. This type of solution, when compared to time-domain simulations, is significantly less costly in terms of the time and computing power required (JR and R, 1985).

This work will use frequency-domain solutions to determine the frequency response of certain structures and then compare these results with those obtained by numerical analysis in the time domain.

3.1.1 Modal Analysis

When you want to know the natural frequencies and vibration modes of a finite element system, you have to solve an eigenvector and eigenvalue problem. For a particular case of a system with no damping and no external force, the equation of motion can be written as follows

$$M\ddot{\vec{U}} + K\vec{U} = 0 \qquad (12)$$

where M is the mass matrix and K is the stiffness matrix, the development of these matrices will be described below. An eigenvector problem for this equation can be written as

$$(K - \lambda M)\vec{U} = 0 \qquad (13)$$

and by manipulating this equation, multiplying both sides of the equation by the inverse of the mass matrix, you get

$$(M^{-1}K - \lambda I)\vec{U} = 0 \qquad (14)$$

This representation of this equation is only illustrative, since inverting matrices is a procedure that is usually avoided due to its high computational cost; The solution of this equation will result in the eigenvalues and their associated eigenvectors.

There is a direct relationship between each eigenvector and the system's natural frequencies, given by

$$\omega_i = \sqrt{\lambda_i} \qquad (15)$$

where ω_i is the i-th natural frequency of the system, associated with the eigenvector λ_i. Each eigenvector has an associated eigenvalue, which also has a Hsical interpretation; this represents the modal form of the system when excited with the natural frequency associated with that eigenvalue.

This relationship will be used in this work to compare time-domain simulations with frequency-domain simulations. The modal shapes will be simulated in the frequency domain and then a time-domain simulation will be used. In the time domain, the frequency of the excitation force will be one of the natural frequencies. The normalised vibration amplitudes will then be compared with the eigenvectors. In short, we want to compare the structure excited at one of the modal frequencies with its corresponding eigenvector.

3.1.2 Euler-Bernoulli beam

The Euler-Bernoulli beam model was chosen for the numerical model. This method was chosen mainly due to its simplicity and ease of implementation. Some of the main assumptions of this model are that the transverse section of the beam is rigid in the plane that contains it, that this section remains flat after deformation and that it remains normal to the deformed axis of the beam (BAUCHAU and CRAIG, 2009). For the model to be valid, only small deformations and a relatively long beam must be contained.

Figure 9 - 2-node beam element

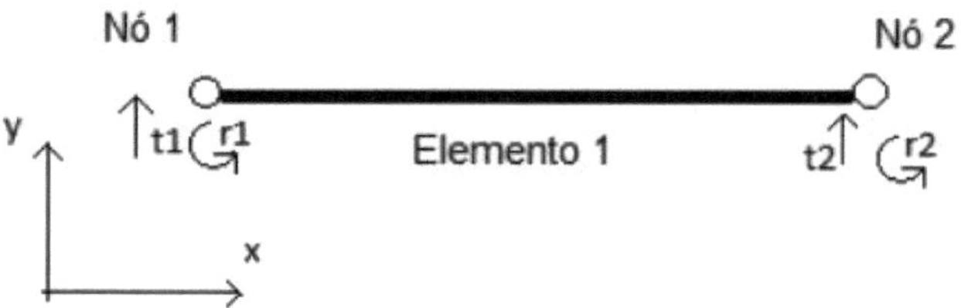

Source: Author himself

The numerical model used a beam element with two nodes, each node having one rotational and one translational degree of freedom, as shown in Figure 9. The material of the model is isotropic and homogeneous. The force functions of this element are given by cubic polynomials as a function of *r*, which is the local coordinate system that varies from -1 in no. 1 to 1 in no. 2, and *le is* the length of the element. The shape functions are given by

$$N_1 = \frac{1}{4}(1-r)^2(2+r) \tag{16}$$

$$N_2 = \frac{1}{8}l_e(1-r)^2(1+r) \tag{17}$$

$$N_3 = \frac{1}{4}(1+r)^2(2-r) \tag{18}$$

$$N_4 = -\frac{1}{8}l_e(1+r)^2(1-r) \tag{19}$$

N1 and *N3* are associated with the translational degrees of freedom of nodes 1 and 2 respectively, and *N2* and *N4* with the rotational degrees of freedom of the same nodes. These four shape functions can be organised in the form of an N^e , 4x1 matrix, where each of the shape functions corresponds to a row. The mass matrix for this element can be obtained through an analytical integration

$$M = \rho A \int_{-1}^{1} J(N^e)^T N^e dr \tag{20}$$

where ρ is the density of the material, *A* is the cross-sectional area and *J* is the matrix

If we solve this integral analytically, we get

$$M = \frac{m^e}{420}\begin{bmatrix} 156 & 22l_e & 54 & -13l_e \\ 22l_e & 4l_e^{\ 2} & 13l_e & -3l_e^{\ 2} \\ 54 & 13l_e & 156 & -22l_e \\ -13l_e & -3l_e^{\ 2} & -22l_e & 4l_e^{\ 2} \end{bmatrix} \tag{21}$$

where m^e is the mass of the element.

The procedure for calculating the stiffness matrix begins with the calculation of a matrix B,

$$B = \frac{d^2N}{dx^2} \tag{22}$$

resulting in

$$B = \frac{1}{l_e}\begin{bmatrix} \frac{6r}{l_e} & 3r-1 & -\frac{6r}{l_e} & 3r+1 \end{bmatrix} \tag{23}$$

Once you have B, you can now calculate the local stiffness matrix, *K,* which is given by

$$K = \rho A \int_{-1}^{1} E * I(B)^T B * \frac{1}{2} * l_e * dr \tag{24}$$

By performing this operation, we get

$$K = EI\begin{bmatrix} \frac{12}{l_e^3} & \frac{6}{l_e^2} & -\frac{12}{l_e^3} & \frac{6}{l_e^2} \\ \frac{6}{l_e^2} & \frac{4}{l_e} & -\frac{6}{l_e^2} & \frac{2}{l_e} \\ -\frac{12}{l_e^3} & -\frac{6}{l_e^2} & \frac{12}{l_e^3} & -\frac{6}{l_e^2} \\ \frac{6}{l_e^2} & \frac{2}{l_e} & -\frac{6}{l_e^2} & \frac{4}{l_e} \end{bmatrix} \quad (25)$$

Although very practical and widely used, certain aspects present in the real situation to be studied cannot be solved by frequency-domain simulations. For example, the variations in mass and stiffness throughout the process, the frequency domain simulation is capable of including such vibrations, and countless simulations would have to be carried out for certain stiffnesses and masses.

Another example is force modelling, which is time-dependent since it depends on the position of each tooth at a given time, as well as depending on the displacement of the handle, for example, if the handle moves towards the tool, a greater amount of material will be removed, and greater amounts of force will be involved. The opposite is also true: at a given time, the grip moves in the opposite direction to the tool, and as a result, a greater amount of material is removed by the tool, and depending on these amplitudes, the engagement between tool and grip can be lost.

In order to remedy these limitations, we sought to use a time-domain simulation, capable of incorporating the aforementioned contact between support and beam into the numerical model, as well as applying a time-varying force model. One of the most important parts of time-domain simulation is the integration methods used (WILSON A. ARTUZI, 2005). The following subchapters will explain the time integration method applied, as well as the assumptions and simplifications made in this model.

In order to perform the integration over time, we chose to use the Hubolt method. This method is unconditionally stable, meaning that even with large time increments, it will not present divergent solutions. Hubolt's integration is also not self-initialising (HUGHES, 1987), i.e. it requires an additional method to calculate one or more initial parameters. This feature of the method will become more evident after the equations have been developed.

Since *U is* the displacement of a given degree of freedom at time *t*, U^* *is* its first time derivative and so on, it follows that for a truncation in the cubic term of the Taylor series

$$U^{t} = U^{t+\Delta t} + (-\Delta t)\dot{U}^{t+\Delta t} + \left(\frac{-\Delta t}{2}\right)^{2} \ddot{U}^{t+\Delta t} + \left(\frac{-\Delta t}{6}\right)^{3} \dddot{U}^{t+\Delta t} \quad (26)$$

$$U^{t-\Delta t} = U^{t+\Delta t} + (-2\Delta t)\dot{U}^{t+\Delta t} + \left(\frac{-2\Delta t}{2}\right)^{2} \ddot{U}^{t+\Delta t} + \left(\frac{-2\Delta t}{6}\right)^{3} \dddot{U}^{t+\Delta t} \quad (27)$$

$$U^{t-2\Delta t} = U^{t+\Delta t} + (-3\Delta t)\dot{U}^{t+\Delta t} + \left(\frac{-3\Delta t}{2}\right)^{2} \ddot{U}^{t+\Delta t} + \left(\frac{-3\Delta t}{6}\right)^{3} \dddot{U}^{t+\Delta t} \quad (28)$$

and manipulating these equations in order to isolate $\ddot{U}^{t+\Delta t}$ e $\dot{U}^{t+\Delta t}$, as

$$\ddot{U}^{t+\Delta t} = \frac{1}{\Delta t^{2}} [2U^{t+\Delta t} - 5U^{t} + 4U^{t-\Delta t} - U^{t-2\Delta t}] \quad (29)$$

$$\dot{U}^{t+\Delta t} = \frac{1}{6\Delta t^{2}} [11U^{t+\Delta t} - 18U^{t} + 9U^{t-\Delta t} - 2U^{t-2\Delta t}] \quad (30)$$

Substitute the values from the equations of $\ddot{U}^{t+\Delta t}$ e $\dot{U}^{t+\Delta t}$ into the equation of motion,

$$M\ddot{U}^{t+\Delta t} + C\dot{U}^{t+\Delta t} + KU^{t+\Delta t} = F^{t+\Delta t} \quad (31)$$

one has

$$\left[\frac{2}{\Delta t^{2}} M + \frac{11}{6\Delta t} C + K\right] U^{t+\Delta t} \quad (32)$$

$$= F^{t+\Delta t} + \left[\frac{5}{\Delta t^{2}} M + \frac{3}{6\Delta t} C\right] U^{t} - \left[\frac{4}{\Delta t^{2}} M + \frac{3}{\Delta t} C\right] U^{t-\Delta t}$$

$$+ [\frac{1}{\Delta t^{2}} M + \frac{1}{3\Delta t} C] U^{t-2\Delta t}$$

However, at the initial moment $U^{t-2\Delta t}$, is unknown, so as mentioned above, this method needs another method to run the first integration loop.

3.2.1 Modelling the Machining Force

The magnitude of the force, as well as its direction, depends on various factors, such as the geometry and number of teeth of the tool, the material of the grip and the machining parameters in general. It is intuitive that the greater the depth of cut, the greater the forces involved. Altintas models the cutting force of each tooth in the time domain, taking into account the position of the cutter tooth at the time (ALTINTAS, 2011).

Figure 10 - Variation in chip thickness during tooth passage

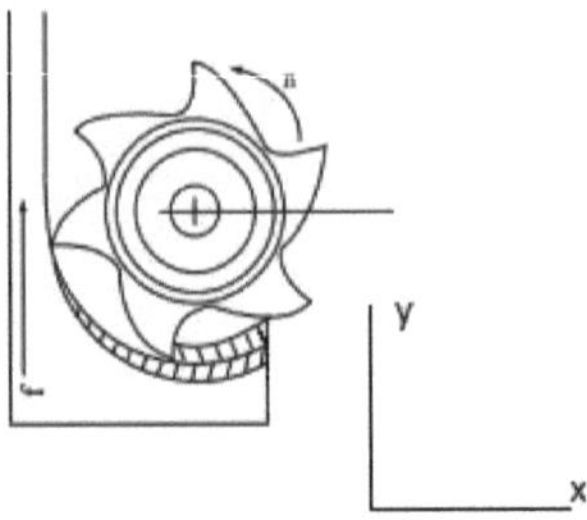

Source: (ALTINTAS, 2011)

Figure 10 shows the variation in chip thickness during the entry and exit of a tooth. During entry, the angle of the tooth, ^at, is considered to be zero, and so is the thickness of the chip, *h.* As the tooth rotates, the thickness increases, reaching its peak when the angle of the tooth is 90°. After this, the chip thickness decreases again until it reaches the exit angle, ^ex. There are situations where the depth of cut is less than the tool radius, in which case the exit angle is less than 90° and this angle is taken as the position of greatest chip thickness.

It is possible to draw up an equation for the thickness of the chip as a function of the feed rate, *c*, the unit of which is m/(revolution*tooth). Figure 11 helps to visualise this mathematical relationship: the circle above represents the tool as it enters a given tooth, and the circle below represents the tool in the position where that tooth is leaving the grip. The tool's trajectory passes through the position of the entering tooth and the position of the exiting tooth. If you make the x-axis parallel to the feed, and the *y-axis* normal to it and in the opposite direction to the handle, you can obtain a relationship. The tool path passes through the points mentioned above and has its derivative with respect to *x* zero at the initial instant. Additional information is that at 90° the chip thickness is equal to the rate of advance (difference in the position of the two circles), so $h(v) = c * sinp$ (33)

where ^ is the angular position at a given instant.

Figure 11 - Tool position at different times during machining

Source: Author himself

However, this chip thickness equation is only valid for fixed handles and tools, i.e. there is no movement between them on the *y-axis*. The present study aims to study flexible handles, and therefore some adjustments can be made to this equation. Assuming that the handle only moves in the *y-axis*, Figure 11 represents these movements. To make it easier to visualise, the coordinate system in this case is considered to be attached to the handle. The rightmost circle represents the tool standing still, but if the tool moves relative to the handle or the other way round, the rightmost circle represents this condition.

It can be seen that the chip thickness would be zero in the first condition at all instants without advancing. However, for the approach movement between the handles, with no feed rate, the chip has a maximum thickness for an angular position equal to zero, and zero thickness for a tooth at 90°. You can therefore write the chip thickness as a function of the handle displacement as

Figure 12 - Relative movement between handle and tool

Source: Author himself

$$h(\varphi) = c * sin\varphi \quad (34)$$

where *y is* the displacement of the handle.

You can combine both equations, resulting in a formulation that takes into account the influence of both displacement and feed rate on the chip thickness for a given tooth position, where the chip thickness is

$$h(\varphi) = y * cos\varphi + c * sin\varphi \quad (35)$$

It is worth noting that the *y* displacement of the handle is approximate; in the finite element code, this displacement is given by an interpolation between the displacement of the two nearest nodes, this interpolation is done using the element's shape functions.

Knowing the thickness of the chip as a function of the angular position of the tooth, (ALTINTAS, 2011) force equations are established, where the components of the tangential, radial and axial forces are respectively

$$F_t(\varphi) = K_{tc}a_c h(\varphi) + K_{te}a_c \quad (36)$$

$$F_t(\varphi) = K_{rc}a_c h(\varphi) + K_{re}a_c \quad (37)$$

$$F_t(\varphi) = K_{ac}a_c h(\varphi) + K_{ae}a_c \quad (38)$$

Ktc, *Krc* and *Kac* are the shear force coefficients that contribute to shear in the aforementioned directions, and *Kte*, *Kre* and *Kae* are shear edge constants (*z-direction*).

Although this coordinate system is very practical for understanding the forces involved, it is necessary to convert them to the global coordinate system. The forces in the global system, in the *x*, *y* and *z* directions, are respectively

$$F_x(\varphi) = -F_t * cos(\varphi) - F_r * sin(\varphi) \quad (39)$$

$$F_y(\varphi) = F_t * sin(\varphi) - F_r * cos(\varphi) \quad (40)$$

$$F_z(\varphi) = F_a \quad (41)$$

It is intuitive that each tooth will only exert force on the workpiece if it is engaged, in other words,

$$F_x(\varphi), F_y(\varphi)\ e\ F_z(\varphi) = 0\ when\ \varphi < \varphi_{st}\ ou\ \varphi < \varphi_{ex} \quad (42)$$

where φst is the entry angle and φex is the exit angle. Therefore, before the tooth enters and after it leaves contact with the workpiece, the forces exerted by it will be zero. The spacing angle between the teeth is given by

$$\varphi_p = \frac{2\pi}{N} \quad (43)$$

with N representing the number of teeth on the cutter.

The resulting force in the direction of the *x*, *y* and *z* axis can be achieved by adding up the forces of each tooth in that direction,

$$F_x(\varphi) = \sum_{j=1}^{N} F_{xj}(\varphi_j) \quad (44)$$

$$F_y(\varphi) = \sum_{j=1}^{N} F_{yj}(\varphi_j) \quad (45)$$

$$F_z(\varphi) = \sum_{j=1}^{N} F_{zj}(\varphi_j) \quad (46)$$

Fxj represents the *x* component of the force on tooth *j*, and the same is true for the other two directions. Remember that if φj is outside the range of entry and exit of the teeth, the resulting force will be zero. It is also known that each tooth is offset from the posterior tooth by φp radians.

In the developed finite element code, the resulting force in a given direction will be between two nodes, and this will be distributed between them by means of an interpolation using the shape functions again.

Knowing the tangential component of the force, Ft, and the diameter of the cutter, *D*, it is possible to estimate the torque on the tool axis, Tc, which is

$$T_c = \frac{D}{2} \cdot \sum_{j=1}^{N} F_{tj}(\varphi_j) \quad (47)$$

This tangential force also provides information on the cutting power, P_c,

$$P_c = V \cdot \sum_{j=1}^{N} F_{tj}(\varphi_j) \quad (48)$$

where *V is* the cutting speed and is given by

$$P_c = V \cdot \sum_{j=1}^{N} F_{tj}(\varphi_j) \quad (49)$$

where *n is* the rotation of the tool axis.

3.3 Hypotheses

The machining operation encompasses various mechanical phenomena such as plasticity, temperature variations, friction and dynamic contact. It is unfeasible and extremely difficult to build a mathematical model capable of encompassing all these aspects and still be solvable in a reasonable amount of time. It is therefore necessary to apply some simplifications.

A first hypothesis concerns the shear forces in the beam, which due to the long beam model are neglected. A perfectly rigid crimp was also applied, in which the displacements are zero. The beam material was also considered to be isotropic, continuous and homogeneous. The beam model applied does not take into account the forces normal to the transverse section of the beam.

With regard to cutting forces, these were summarised as tangential, normal and axial forces on the tool, and contact and plasticity effects were summarised in a much more simplified force model. Temperature effects in the cutting region, as well as friction and contact are also not computed.

4 RESULTS AND DISCUSSIONS

All the results presented here were generated using our own computer code in Scilab. The results of the numerical simulations, as well as the mesh refinement analysis, will be presented and discussed in this chapter.

4.1 Mesh Refining Analysis

The refinement analysis is a fundamental stage in all finite element simulations. It tells us whether the model can provide an accurate response with a given number of elements. In this study, the first two natural frequencies of a beam were used as reference parameters. Table 1 shows the analytical natural frequencies for a braced beam.

Table 1 - Analytical natural frequencies of a braced beam

Mode	Fn (Hz)	Wn (rad/s)
1°	331	2081
2°	2076	13044
3°	5813	36527

Source: Author

Figure 13 and Figure 14 show the first and second natural frequencies obtained numerically for the beams, respectively. As expected, the convergence of the second mode is slower, since as it is a more complex vibration mode than the first, it needs more elements to be correctly represented.

Figure 13 - Convergence analysis, 1st natural frequency

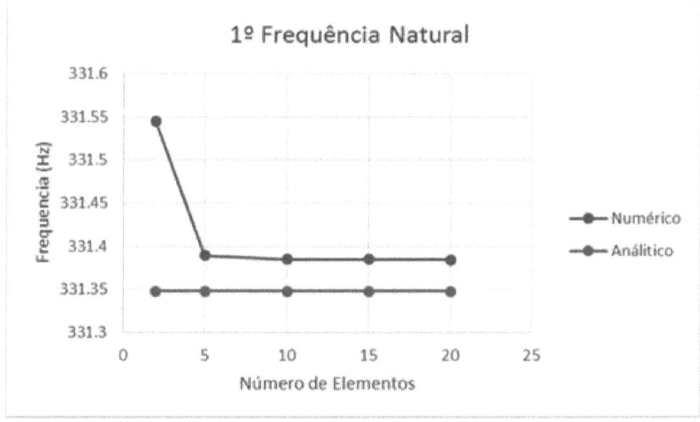

Source: Author

Figure 14 - Convergence analysis, 2nd natural frequency

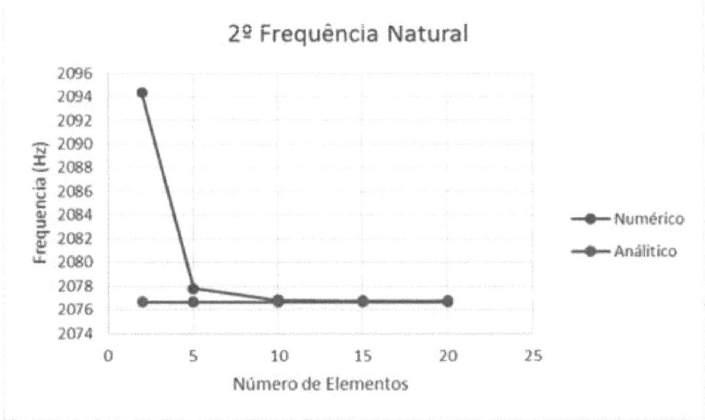

Source: Author

It was observed that 10 elements are enough to represent the first two modes correctly. It may be questioned why only the first two modes are taken into account. This is due to the fact that for the geometry studied, the third mode already has a very high frequency, which is practically not excited during a machining operation, and therefore does not need to be taken into account. It should be noted that this hypothesis is only valid for this geometry and process, and similar hypotheses should be studied on a case-by-case basis.

4.2 Frequency Domain Simulation

A cast, rectangular beam was simulated, the properties of which are described in Table 2. Figure 15 shows the first 4 vibration modes. These results will later be compared with those obtained by time-domain simulation, and will serve as one of the validation methods for this numerical model.

Table 2 - Geometrical, material and mesh properties

Type of support	Crimping
Young's Modulus (GPa)	210
Beam length (m)	0,2
Beam height (m)	0,016
Beam width (m)	0,016
Number of elements	20
Density (Kg/m)^3	7860

Source: Author

Figure 15 - Vibration Modes

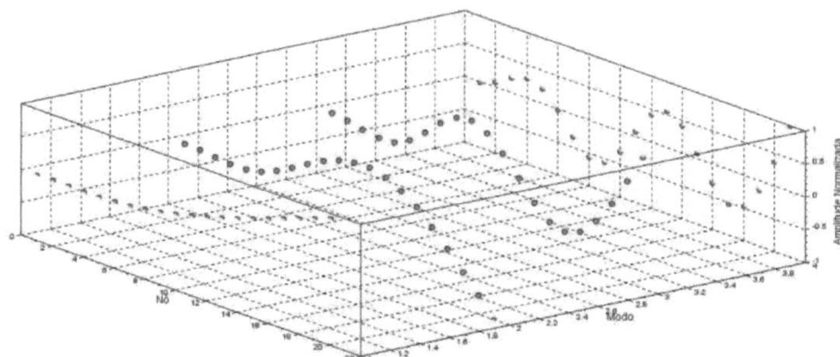

Still in the frequency domain, the vibration modes were simulated for the beam at different time instants (as the cutter advances, the beam has more material being removed). This simulation used the aforementioned properties from Table 2 and, in addition, the properties from Table 3. The aim of these simulations is to discover the dynamic behaviour of the beam at different times during machining, in order to represent the effect of material removal. In other words, as the cutter advances, an amount of material is removed and the dynamic characteristics of the structure are modified. This is basically due to the fact that the stiffness and mass matrices of the elements already machined are altered. Figure 16 shows the first vibration modes of 10 structures, with structure number 10 without any material removal, and in descending order on the z-axis there are structures with an incremental material removal of 0.01 m, up to structure 1, which has half its length machined.

Table 3 - Structure properties

Number of structures	10
Incremental distance (m)	0,01
Depth of material removed (m)	0,005

Source: Author himself

Figure 16 - First vibration mode for various structures

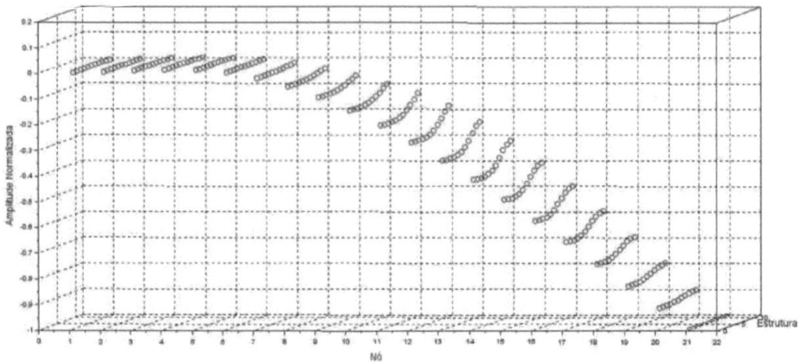

4.3 Computer Code Validation

In order to validate the code implemented for the time-domain simulation, the results obtained from this model were compared with those from the time-domain. From the solution of eigenvectors and eigenvalues of the frequency-domain simulation, we extracted the natural frequencies and the behaviour of the structure for that given frequency. Once we had this information, we used the natural frequencies as the frequencies of the excitation force, which was placed at the free end. The simulation was then run several times, once for each excitation frequency coinciding with the natural frequencies. Then, for a given time, the vibration amplitudes in the nodes were read, normalising the response. Finally, these values were compared with those in the frequency domain for each of the 4 natural frequencies. Figure 17 and Figure 18 show the comparisons for the first and second natural frequencies respectively.

Figure 17 - Comparison between time and frequency domain analysis for the first vibration mode

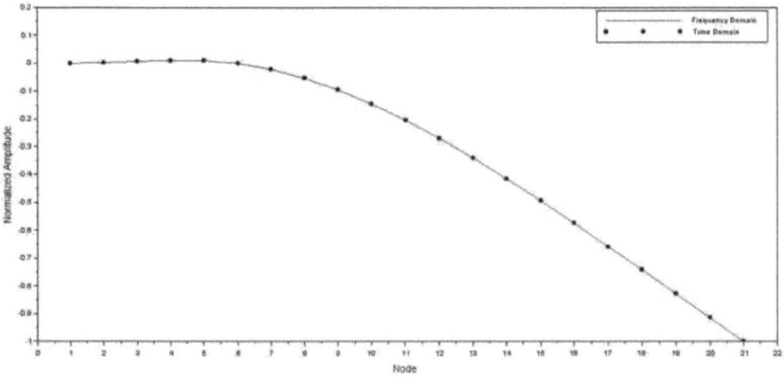

Source: Author himself

Figure 18 - Comparison between time and frequency domain analysis for the second

vibration mode

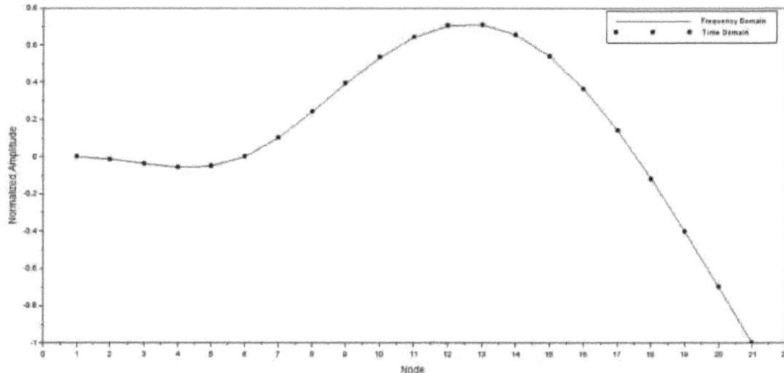

Source: Author himself

It can be seen that both simulations are in agreement. Normalising the amplitudes is essential, as it makes it possible to compare the vibration modes of the curves between the time and frequency domains.

4.4 Time Domain Simulation

Finally, time-domain simulation was used. The Altintas forging model was used, with adaptations for flexible grip, as described in subchapter 3.2.1 - Machining Forging Modelling. The beam model used was the Euler-Bernoulli model and time-domain integration was carried out using the Hubolt method. The data used in the simulation, such as material properties, machining parameters and coefficients, number of increments and process time are illustrated in Table 4.

The parameters were chosen on the basis of a real machining situation for an auger beam. The K_{rc} and K_{re} coefficients were used based on references in the literature (POWELL, 2008), but as they depend on the parameters of the tool and the grip, they must be measured for each specific situation.

Table 4 - Frequency domain simulation data

Number of us	11
Beam length (m)	0,2
Beam height (m)	0,016
Beam width (m)	0,016
Density (Kg)	7860
Number of teeth	4
Cutter speed (rpm)	1200
Avango speed (m/s)	0,02

Tool radius (m)	0,005
End time (s)	5
Number of increments	100000
Krc(N/m2)	10000
Kre(N/m)	0
Depth of cut (m)	0,005

Source: Author himself

Figure 19 shows the force applied to the translational degrees of freedom, with the dark blue being the translational degree of freedom of the free end, the red the previous one and so on. These five degrees of freedom were chosen for a reason: they are where the cutter is passing. From the data shown in Table 4, the cutter takes 1 second to travel 0.1 metres, which is exactly the distance between us. Therefore, after a time of 0 seconds, the force component will be maximum at the free end point, and after 1 second it will be maximum at the previous point, as shown in the graph.

Figure 19 - Load on each degree of freedom over time

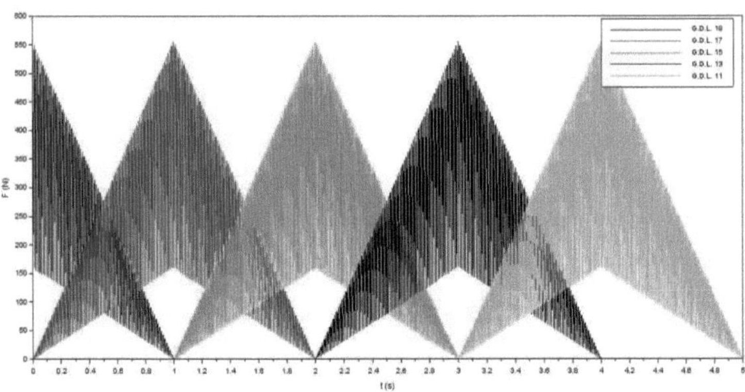

Source: Author himself

Figure 20 is just an approximation of the previous image, and is intended to illustrate how the code weights the cutting forces between the two nodes. At the initial time of the graph, it can be seen that the force is closer to the end node (blue) and is therefore greater there. As the cutter moves, it gets closer to the next node and further away from the end, so it becomes greater at the second-to-last node (red).

Figure 20 - Load on the last two translation degrees of freedom

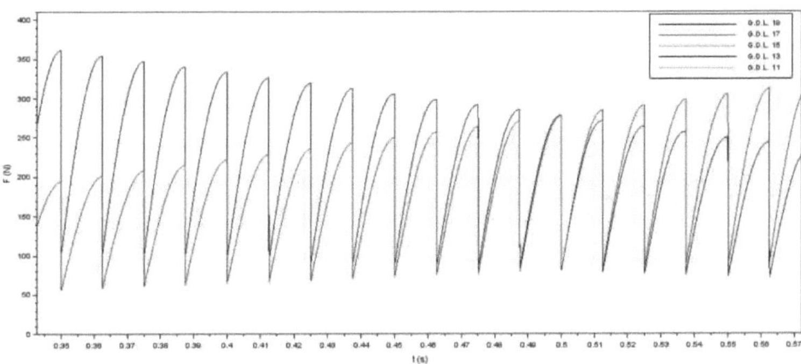

Source: Author himself

Figure 21 represents the total machining force. It has the characteristic sawtooth shape (AYKUT, BAGCIB, *et al.,* 2006). The physical interpretation is simple: the tooth comes into contact with the handle, and as the tool rotates, the chip thickness increases, consequently the cutting force increases, until the tooth leaves contact with the handle. This phenomenon is in accordance with Equation (33) (ALTINTAS, 2011).

Figure 21 - Total machining force

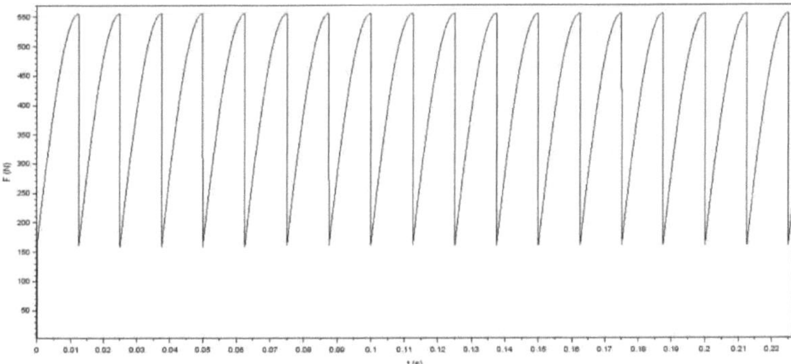

Source: Author himself

Finally, the displacements in the different degrees of freedom can be analysed. Figure 22 illustrates the displacements in the nodes where the cutter is passing over time. It can be seen that as the cutter moves away from the free end, it has a smaller moment arm, and as a result the displacement amplitudes decrease over time.

Figure 22 - Displacements of the nodes over time

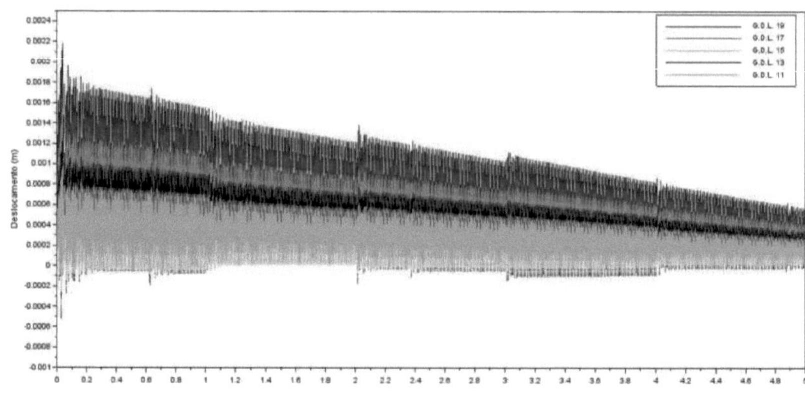

t(S)

Source: Author himself

However, in order to study the surface finish and roughness, only the displacement in the region where the cutter is is important, since this is where the removal of material occurs at a given instant of time. Figure 23 illustrates the displacements in the places where the cutter is passing at each instant. It is known that there are not enough nodes for each position of the cutter in time (you would need a node for each time increment, which is unfeasible), so when the cutter is between two nodes, the desired displacement is weighted by the displacements of the nearest nodes.

Figure 23 - Displacements in the cutter contact region over time

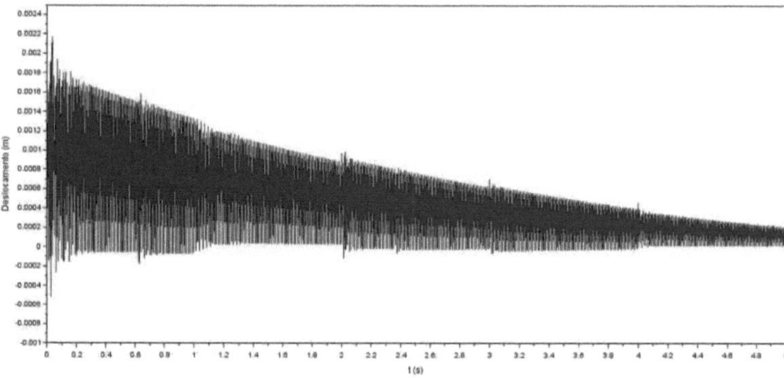

Source: Author

It is clear that the highest vibration amplitudes are present at the initial moments of machining, and as a result, the worst surface finishes. This response corresponds to what is expected by the phenomenon, since as the cutter is closer to the free end, it has a greater lever arm and causes greater bending moments.

It should be noted that this numerical simulation applied a radial machining depth, ar, that was too large. This was done in order to check how the code reacts to large variations in the rigidity of the elements over time. In a real machining situation, the values of this depth rarely exceed 1 miKometre, and in the case of the simulation we used 5 mm.

4.5 Experimental Procedure and Comparisons

A milling process was carried out on three samples, all with the same machining parameters, which are shown in Table 5. These parameters were chosen according to the tool manufacturer's recommendations, combined with the torque and power capabilities of the machine tool. Time-domain simulations (as in the previous section) were carried out for these new parameters.

Table 5 - Experimental machining parameters

Tool radius (m)	0,012
Number of teeth	4
Avango per tooth (mm/tooth)	0,083
Avango (m/s)	0,02
Speed (rpm)	4000
Radial depth of cut (m)	0,001

Source: Author himself

The handles were milled and then their surface finish was assessed using roughness measurements. Due to equipment limitations, roughness was the only parameter measured experimentally. Amplitude measurements during the milling process would provide extremely useful information; such measurements could be made using a laser vibrometer, for example.

A Mitutoyo SJ-210 rugosimeter was used to measure roughness, and the average roughness (*Ra*) was measured for each of the milled samples. The measurement length was 0.02 metres and the positions are shown in Figure 24. Roughness *Ra1* is measured at the free end, while *Ra5* is measured near the middle of the beam. These five roughnesses cover the entire region through which the tool has machined the handle.

Figure 24 - Roughness measurement positions

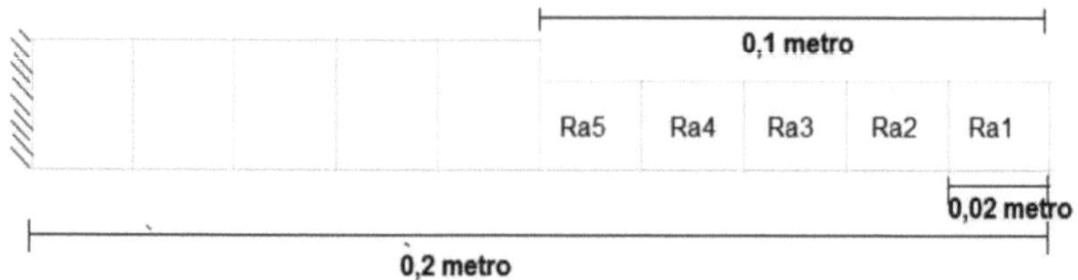

Source: Author himself

Table 6 shows the measured roughness of the three samples in different positions. It can be seen that, as expected, the greatest roughness is in the regions close to the free end, where the greatest vibration amplitudes are concentrated.

Table 6 - Average roughness of the 3 samples

Average roughness (pm)	Sample 1	Sample 2	Sample 3
Ra1	4.18	4.04	5.09
Ra2	2.65	2.52	3.19
Ra3	2.01	2.13	2.75
Ra4	2.00	1.77	2.31
Ra5	1.12	1.24	1.38

Source: Author himself

Finally, these roughness values were compared with the vibrations of the grip at the point where the cutter passed. According to the aforementioned literature, a direct correlation between these two parameters is expected, i.e. higher vibration amplitudes correspond to higher roughness values.

Figure 25 - Average experimental roughness and numerical vibration amplitudes

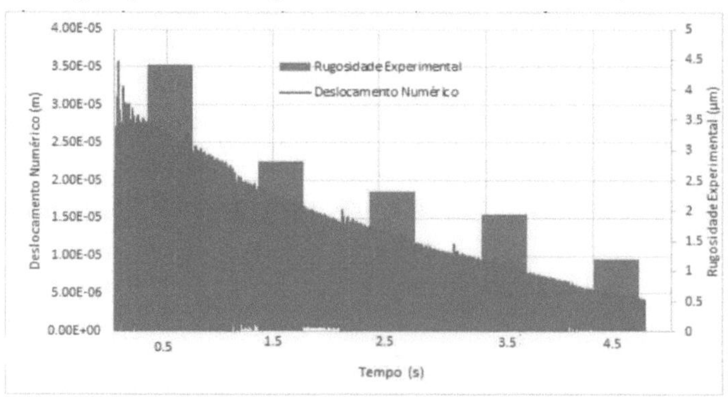

Source: Author himself

Figure 25 shows the vibration amplitudes of the cutter in blue, while the other series of data refers to the average of the three roughness measurements, one average for each vibration region. It can be seen that the predicted behaviour was verified, i.e. the free end of the beam with the highest vibration amplitudes was the same region with the highest roughness. However, although there is a correlation between average roughness and displacement amplitude, the behaviour is not exactly linear. The average roughness *Ra1* was much higher than the others; it was expected that this would be the highest, but such a difference with the others was not expected. One of the reasons for this may be that the Ra1 reading region is where the cutter enters the grip, causing certain initial instabilities that the numerical model is unable to predict. However, a more detailed analysis is required for this specific case, and here again the reading of displacements during the process could provide more information.

5 CONCLUSIONS

At first glance, the results of the time-domain simulations were consistent with those of the frequency domain, with the natural frequencies and vibration modes being extremely close. However, some of the assumptions used in the time-domain simulations cannot be validated in the frequency domain, the main ones being the force model used and the gradual variations in the mass and stiffness of the beam. Both models could be validated by an experimental procedure, as mentioned above.

Hubolt's time integration model, as well as the force model, proved to be very coherent. The force profile found in the simulation was "sawtooth", which corresponds to the models proposed in the literature, as well as to the theoretical reference model (ALTINTAS, 2000).

The computer code was well optimised, being able to run simulations with more than five hundred thousand time increments, which is a very high value if you take into account that with each increment, you have to recalculate the mass and stiffness matrices, as well as the force vector, and solve the linear system for the displacement.

Vibration amplitudes and grip displacements during the numerical simulation were shown to be directly related to surface finish. Places with large vibration amplitudes at the moment the cutter passes showed greater roughness. Therefore, the numerical model is able to indicate where the worst surface finishes will occur for a given machining condition, allowing the processor to take action to reduce these roughnesses. For example, the rotation of the cutter can be altered at points of greatest excitation in order to reduce vibration amplitudes.

It was observed that, once the basic phenomenon is known, it is possible to make simplifications to the model which make it infinitely simpler than the real problem, yet still able to provide relevant information about the process and its parameters. These simplifications have enabled the implemented code to achieve short solving times for the finite element problem.

5.1 Suggestions for future work

It is well known that every numerical model must be validated through experiments, so the first complementary step in this study would be to machine components, accompanied by vibration readings during the process, using a vibrometer for example. Such a procedure could either validate the numerical model, or indicate some hypothetical flaw in the proposal.

Some coefficients would need to be defined, such as the coefficients involved in the cutting force model, but these are only obtained experimentally. They would be interesting for calculating the power and torque required for the machine, as well as reading the amplitudes of the displacements caused by these forces. Although this is not a vital step, since using unit coefficients we can capture the tendencies of the structure's response to a given force.

Finally, as this is an introductory work and has a relatively short development time, this work only deals with beam elements. Analyses of processes in other geometries, such as plates or even three-dimensional solids, would be extremely useful.

Complementary studies for machining excitations close to some of the beam's natural frequencies would also be interesting, since for such operating ranges, the vibration amplitudes are much higher and, as a result, surface finish would be an extremely critical factor.

BIBLIOGRAPHY

ALTINTAS, Y. **Manufacturing Automation:** Metal Cutting Mechanics, Machine Tool Vibrations, and CNC Design. New York: Cambridge University Press, 2000.

ALTINTAS, Y. **Manufacturing Automation**. Vancouver: Cambrudge University Press, 2011.

ALVES, P. R. G. **Analysis of surface finish and power consumed in the machining of eucalyptus wood by tangential front and cylindrical milling**. UNESP. [S.l.]. 2016.

ASADA, H. **Kinematics analysis of workpart fixturingfor flexible assembly with automatically reconfigurable fixtures.** Trans Rob Automat, p. 86-93, 1985.

ASANTE, J. N. **A combined contact elasticity and finite element-based modelfor contact load and pressure distribution calculationin a frictional workpiecefixture system.** The International Journal of Advanced Manufacturing Technology, p. 578, 2008.

AYKUT, S. et al. **Experimental observation of tool wear, cutting forces andchip morphology in face milling of cobalt based super-alloywith physical vapour deposition coated and uncoated tool.** Material and Design, v. 28, p. 1880-1888, 2006.

BAKERJIAN, R. **Tool and manufacturing engineers handbook.** [S.l.]: Society of Manufacturing Engineering , 1992.

BAUCHAU, O. A.; CRAIG, J. I. **Structural Analysis.** [S.l.]: Springer, 2009.

BOYLE, I.; RONG, Y.; BROWN, D. C. **A review and analysis of current computer-aided fixture design approaches.** Robotics and Computer-Integrated Manufacturing, p. 1-12, 2011.

CREDE, C. E.; HARRIS, C. M. **Shock and Vibration Handbook**. [S.l.]: Mc. Graw-Hill, 1961.

DINIZ, A. E.; MARCONDES, F. C.; COPPINI, N. L. **Tecnologia da machinagem dos materiais**. Sao Paulo: MM, 1999.

DOMINGOS, D. C. **Numerical simulation of stresses in the external milling process of trunnions and crankshafts**. Rio de Janeiro: [s.n.], 2002.

DROZDA, T. J.; WICK, C. **Tool and Manufacturing Engineers Handbook - Machining.** Dearborn, Michingan: Society of Manufacturing Engineers, 1983.

EWINS, D. J. **Modal Testing Theory and Practice**. [S.l.]: Research Studies Press, 1984.

FERRARESI, D. **Fundamentals of metal machining**. Sao Paulo: Blucher, 1977.

GRZESIK, W. **Advanced machining process of metallic materials:** theory modelling and applications. Amsterdam: Elsevier, 2008.

HANSEN, M. H. **Aeroelastic stability analysis of wind turbines using an eigenvalue approach.** Wind Energy, v. 7, n. 2, p. 133-143, June 2004.

HUGHES, T. J. R. **Linear Static and Dynamic FInite Element Analysis**. Englewwod Cliffs: Prentice-Hall, 1987.

JAE-WOONG YOUN, Y. J. A. S. P. **Interference-free tool path generation in five-axis machining of a marine propeller.** International Journal of Production Research, v. 41, n. 18, 2003.

JOHNSON, K. **Contact mechanics.** Cambridge University Press, 1985.

JR, C.; R, R. **A review of time-domain and frequency-domain component mode synthesis method**. Joint Mechanics Conference. Albuquerque: Texas University. 1985. p. 1-30.

KONIG, W.; KLOCKE, F. **Fertigungsverfahren, Band 1: Drehen, Frasen, Bohren.** Berlin: Springer, 1999.

KRATOCHVIL, R. **FINISH MILLING AT HIGH CUTTING SPEEDS FOR INDUSTRIAL GRAPHITE ELECTRODES**. FEDERAL UNIVERSITY OF SANTA. Florianopolis, p. 119. 2004.

KRISHNAKUMAR, K.; MELKOTE, S. N. **Machining fixture layout optimisation using the genetic algorithm.** International Journal of Machine Tools and Manufacture, p. 579-598, 2000.

KUO, H.-C.; DZAN, W.-Y. **The analysis of NC machining efficiency for marine propellers.** Journal of Materials Processing Technology, v. 124, n. 3, p. 389-395, June 2002.

LEE, E.; NIAN, C.; TARG Y. **Design of a dynamic vibration absorber against vibrations in turning operations.** Journal of Materials Processing Technology, p. 278-285, 2001.

M. JURECZKO, M. P. A. M. **Optimisation of wind turbine blades.** Journal of Materials Processing Technology , p. 463-471, 2005.

NCG. **Testing Guidelines and Testing Workpieces for High Speed Cutting**. [S.l.]: NC-Gesellschaf, 2000.

PIMENTEL, R. **Modelling and simulation of machining processes.** Seminar on Machining Technology, Sao Paulo, 2011.

POWELL, K. B. **CUTTING PERFORMANCE AND STABILITY OF HELICAL ENDMILLS WITH VARIABLE PITCH**. UNIVERSITY OF FLORIDA. [S.l.]. 2008.

RIVIN, E. I.; KANG, H. **Improvement of machining conditions for slender parts by tuned dynamic stiffness of tool**. International Journal of Machine Tools and Manufacture, p. 361-376, 1989.

SCHMITZ, T. L.; DONALSON, R. R. **Predicting High-speed Machining Dynamics by Substructure Analysis**, 2000. 303-308.

SCHUKZ, H.; WURZ, T.; BOHNER, S. **Proper tool balancing.** Maquinas e Metais, p. 24-31, 2001.

SMITH, S. S.; TLUSTY, J. J. **Update on High-Speed Milling Dynamics.** ASME, p. 142-149, 1990.

STEMMER, C. **Cutting tools I**. Florianopolis: UFSC, 1995.

STOETERAUS, L. M. **Static and dynamic simulation of a CNC lathe for ultra-precision machining.** XV Brazilian Congress of Mechanical Engineering, Rio de Janeiro, 1999.

TARNG, Y. S.; KAO, J. Y.; LEE E.C. **Chatter suppression in turning operations with a tuned vibration absorber.** Journal of Materials Processing Technology, p. 5560, 2000.

TLUSTY, G. **Manufacturing Process and Equipment**. New Jersey: Prentice Hall, 2000.

TOBIAS, S. A. **Machine Tool Vibration**. [S.l.]: Blackie and Sons, 1965.

WERNER, A. **ProzeBauslegung und ProzeBsicherheit beim Einsatz von schlanken Schaftfrasern.** [S.l.]: [s.n.], 1992.

WILSON A. ARTUZI, J. **Improving the Newmark Time Integration Scheme in Finite Element Time Domain Method.** IEEE MICROWAVE AND WIRELESS COMPONENTS LETTERS, v. 15, n. 12, December 2005.

WIPPLINGER, K. P. M. . T. M. H. A. A. B. T. **Stainless steel finned tube heat exchanger design for waste heat recovery.** Journal of Energy in Southern Africa, v. 17, n. 2, p. 47-56, 2006.

XIAO, M.; KARUBE, S. . S. T.; SATO, K. **Analysis of chatter suppression in vibration cutting.** International Journal of Machine Tools and Manufacture, p. 16771685, 2002.

XIONG, L.; MOLFINO, R.; ZOPPI, M. **Fixture layout optimisation for flexible aerospace parts based on self-reconfigurable swarm intelligent fixture system.** The International Journal of Advanced Manufacturing Technology, p. 1305-1313, 2013.

ZATARAIN, M. et al. **Stability of milling processes with continuous spindle**

speed variation: analysis in the frequency and time domains, and experimental correlation. CIRP Annals-Manufacturing Technology, 2008.
ZHONGQUN, L.; QIANG, L. **Solution and Analysis of Chatter Stability for End Milling in the Time-domain**. Chinese Journal of Aeronautics, 2008.

APPENDIX

APPENDIX A - SCILAB COMPUTER CODE

```
clear
close
clc
//###########entry data#########
//number of elements
nelems=10;
//number of us
nnos=nelems+1;
//total beam length (m)
Lt=.2;
// Young's modulus (Pa)
E=210e9;
//density (Kg m'~3)
rho=7860;
//transverse beam section width
b=.015875;
//height of the transverse section of the beam
h=.015875;
//h2: thickness after machining h2=0.01;
// time entries
// n is the number of steps
//start time (s)
t0 = 0
//end time (s)
tf =5
//number of steps
n = 100000
//increment of time (s)
dt = (tf-t0)/n
//the cutting depth
ar=h-h2
//##tool parameters and cutting conditions
//rf and tool radius ##change later
rf=air
//nd is the number of teeth on the tool
nd=4
Krc=10" 8*10" -6
Kre=10" 4*10" -6
//avango m/s, ######then make one with avango per tooth####### av=.02
nrpm=1200
//tool frequency
ff=nrpm/60
// tooth by tooth
avd=av/(nd*ff)
```

```
I=b*^3/12;
12=Ь*Ь2л3/12;
A=b*h;
A2=b*h2;
//before machining
prop=[E,I,rho,A];
//after machining
prop2=[E,I2,rho,A2];
zeta1=.1
zeta3=.1
//length of each element
L=Lt/nelems;
//restricted degrees of freedom gl_restricted=[1;2];
//######### End of Input data ############
//beam grinder2
//creates the coordinates of each node
coord=zeros(size(nnos))
for i=1:nnos-1
coord(i+1)=coord(i)+Lt/nelems
//coord(i+nelems+1)=(coord(i+1)+coord(i))/2
end
//matrix indicating the nodes connecting each element
connectivity=zeros(nelems,2)
for i=1:nelems
for j=1:nnos
connectivity(i,1)=i
connectivity(i,2)=i+1
end
end
//Table with degrees of freedom for each beam
EFT=zeros(nelems,4);
for i=1:nelems
no1=connectivity(i,1);
no2=connectivity(i,2);
EFT(i,1)=2*no1-1;
EFT(i,2)=2*no1;
EFT(i,3)=2*no2-1;
EFT(i,4)=2*no2;
end
//function to calculate global beam stiffness
function [K]=rigidity(E, I, L)
K=[(12*E *I)/L^ 3,(6 *E *I)/L^ 2,-(12*E *I)/L^ 3 ,(6*E *I)/L^ 2;
(6 *E *I)/LA2,(4 *E *I)/L,-(6 *E *I)/LA2,(2 *E *I)/L;
-(12*E*I)/LA3,-(6*E*I)/LA2,(12*E*I)/LA3,-(6*E*I)/LA2;
(6*E*I)/LA2,(2*E*I)/L,-(6*E*I)/LA2,(4*E*I)/L]
endfunction
function [M]=mass(rho, A, L)
```

```
M=[(13*rho*A*L)/35,(11*rho*A*LA2)/210,(9*rho*A*L)/70,-(13*rho*A*LA2)/420;
(11*rho*A*LA2)/210,(rho*A*LA3)/105,(13*rho*A*LA2)/420,-(rho*A*LA3)/140;
(9*rho*A*L)/70,(13*rho*A*LA2)/420,(13*rho*A*L)/35,-(11*rho*A*LA2)/210;
-(13*rho*A*LA2)/420,-(rho*A*LA3)/140,-(11*rho*A*LA2)/210,(rho*A*LA3)/105] endfunction
//creating gls
gls = [1:2*nnos]; //Where 2 is the number of degrees of freedom per no.
//Loop to find the boundary conditions:
for i=1:2;
no = 1; //Matches the no with the applied boundary condition.
gl = gl_restricted(i, ); 'Find the gl with the boundary condition applied.
// Zero the gl where a boundary condition is applied:
gls(2*(no-1)+gl) = 0; // Where 2 is the number of degrees of freedom per no.
end; // i
//Matrix of zeros to allocate the global stiffness matrix of all beams
Kaux=zeros(4,4,nelems);
Maux=zeros(4,4,nelems);
Kaux2=zeros(4,4,nelems);
Maux2=zeros(4,4,nelems);
//mass and local stiffness of the beam without machining
Kaux=rigidity(prop( 1,1),prop( 1,2),L);
Maux=massa(prop( 1,3),prop( 1,4),L);
//mass and local stiffness of the machined beam
Kaux2=rigidity(prop2( 1,1),prop2(1,2),L);
Maux2=massa(prop2( 1,3),prop2( 1,4),L);
// initial boundary conditions
U0 = zeros(2*nnos-size(gl_restricted,1),1)
V0 = zeros(2*nnos-size(gl_restricted,1),1)
//forge function
function [F]=forca(t, posi, low, high, Ut_houb, avd, nd, phi, ff, b, rf, ar, Krc, Kre)
F = zeros(2*nnos-size(gl_restricted,1),1)
// strength has to go
//avd=c=avango/dente.rotagao=avango (m/s)/(ndente*rotagao(Wn))
//nd=number of teeth
//phi=Angle
//nrpm=rpm speed
//ff=rotation frequency
//rf=tool radius
//ar=radial depth =height before-width after machining
//he= handle displacement (weighted average between the 2 nos)
he= high*Ut_houb(posi-2,1) + low*Ut_houb(posi,1)
if he>0 then
he=0
end
//chip thickness at the position of each tooth
h=zeros(nd)
//input angle and output angle
//if the angle is between the phie and the phis, the tooth is engaged, as a result we will have force executed by this tooth
```

```
phie=zeros(nd)
phis=zeros(nd)
//Fd and the strength in each tooth
Fd=zeros(nd)
for z=1:nd
phie(z)=(z-1)*2*%pi/nd
phis(z)=phie(z)+%pi/2-asin((rf-ar)/rf)
h(z)=avd*sin(phi-phie(z))+he*cos(phi-phie(z))
if phi>=phie(z) then
if phi<=phis(z) then
Fd(z)=Krc*b*h(z)+Kre*b
end
end
end
Fr = sum(Fd)
F(posi,1)=high*Fr
F(posi+2,1)=low*Fr
endfunction // end function forca
//I'm going to say that the dt has to be at least 5 times shorter than the period of the forga/5
//dt critical
if dt>(2*%pi/(5*nrpm)) then
mprintf('increase the number of increments')
end
//I need to calculate this DFC first, for now I'm setting it to 0
U_dfc=zeros(2*nnos-size(gl_restricted,1),3);
// Inputs to the Houbolt loop
// Reset the time count vector to zero
T = zeros(1,n+1)
Force = zeros(2*nnos-size(gl_restricted,1),n+1)
Phi=zeros(1,n+1)
// Initial Conditions
Umm_houb = U0 // displacement in t-2dt
Um_houb = U_dfc(:,2) // t-dt displacement
Ut_houb = U_dfc(:,3) // displacement in t present
// Zero the displacement vector
U_houb = zeros(2*nnos-size(gl_restricted,1),n+1)
// Reset auxiliary variable
V = zeros(2*nnos-size(gl_restricted,1),3)
// Informs that the force calculation is in the future tense
T(1) = t0
T(2) = t0+dt
T(3) = t0+2*dt
// ########Loop Houbolt##########
t = t0+3*dt
for i = 3:n
//number of shaft turns
nvolt=floor(t*ff)
```

```
phi=2*%pi*t*ff
if phi>2*%pi then
phi=phi-nvolt*2*%pi
end
Phi(i)=phi
// in every loop, I have to zero and then calculate F, M and K, so far I'm only calculating F
//calculation of the no. where the ##forge## is located
//distance travelled
dist=t*av;
//posigao
pos=Lt-dist;
npos=nelems*pos/Lt;
//whole part
ipos=floor(npos);
posi=ipos*2-1
//low and high are the weights of the force in each of the 2 nodes it is between
//decimal part
low=npos-ipos;
//complementary part of decimal
high=1-low;
//Creates the 4 by 4 stiffness matrix
Kglobal=zeros(2*nnos,2*nnos);
Mglobal=zeros(2*nnos,2*nnos);
//allocates the stiffness of each element in the system's stiffness matrix
for r=1:nelems
for q=1:4
for p=1:4
s=EFT(r,q);
ta=EFT(r,p);
if r<= npos then
Kglobal(s,ta)=Kglobal(s,ta)+Kaux(q,p)
Mglobal(s,ta)=Mglobal(s,ta)+Maux(q,p)
else
Kglobal(s,ta)=Kglobal(s,ta)+Kaux2(q,p)
Mglobal(s,ta)=Mglobal(s,ta)+Maux2(q,p)
end
end
end
end
//Applying the boundary conditions to the global stiffness matrix and the force vector
//nrestr=size(gl_restritos,1);
cont = 0 // Initialises the counter.
for j=1:size(gls,2);
if gls(j)==0 then; //If equal to zero, there is a boundary condition applied:
// Delete the row and column of the restricted grade.
Kglobal(:,j-cont) = [];
Kglobal(j-cont,:) = [];
```

```
Mglobal(:,j-cont) = [];
Mglobal(j-cont,:) = [];
cont = cont+1;
end // if.
end // i.
M=Mglobal;
K=Kglobal ;
// Calculate UM_houb (future displacement)
LEFT = ((2*M)/(dt**2)) + K;
RIGHT = forca(t,posi,low,high,Ut houb,avd,nd,phi,ff,b,rf,ar,Krc,Kre) + (((5 *M)/(dt**2)))*Ut_houb -
(((4*M)/(dt**2)))*Um_houb + ((M/(dt**2)) )*Umm_houb;
UM_houb = LEFT\RIGHT;
// Update the variables
V(:,1) = Um_houb;
V(:,2) = Ut_houb;
V(:,3) = UM_houb;
Umm_houb = V(:,1);
Um_houb = V(:,2);
Ut_houb = V(:,3);
// Stores the displacements
U_houb(:,i+1) = Ut_houb;
// Stores the time
T(i+1) = t;
// Stores the forge
Forca(: ,i+1) = forca(t,posi,low,high,Ut houb,avd,nd,phi,ff,b,rf,ar,Krc,Kre);
// Update the time
t = t + dt;
end // end
//######End of Loop######
//Storage of trapezium displacements for plotting
U1_houb = U_houb(2*nnos-1-size(gl_restricted,1),:);
U2_houb = U_houb(2*nnos-3-size(gl_restricted,1),:);
U3_houb = U_houb(2*nnos-5-size(gl_restricted,1),:);
U4_houb = U_houb(2*nnos-7-size(gl_restritos,1),:);
U5_houb = U_houb(2*nnos-9-size(gl_restritos,1),:);
F1_houb = Forca(2*nnos-1-size(gl_restritos,1),:)
F2_houb = Forca(2*nnos-3-size(gl_restritos,1),:)
F3_houb = Forca(2*nnos-5-size(gl_restritos,1),:)
F4_houb = Forca(2*nnos-7-size(gl_restritos,1),:)
F5_houb = Forca(2*nnos-9-size(gl_restritos,1),:)
plot(T,U1 houb, blue")
plot(T,U2 houb,"red")
plot(T,U3 houb,"green")
plot(T.U4 houb,"black")
plot(T,U5 houb,"cyan")
plot(T,F1 houb,"blue")
plot(T,F2 houb,"red")
```

```
plot(T,F3 houb,"green")
plot(T,F4 houb,"black")
plot(T,F5 houb,"cyan")
aaaa=legend(['G.D.L. 19';'G.D.L. 17';'G.D.L. 15';'G.D.L. 13';'G.D.L. 11'])
//there are some frequency-domain signals, for eventual
ffta=f (U1_houb);
// plotffta)
fftb=fft(U2_houb);
// plotffb)
```

yes

I want morebooks!

Buy your books fast and straightforward online - at one of world's fastest growing online book stores! Environmentally sound due to Print-on-Demand technologies.

Buy your books online at

www.morebooks.shop

Kaufen Sie Ihre Bücher schnell und unkompliziert online – auf einer der am schnellsten wachsenden Buchhandelsplattformen weltweit! Dank Print-On-Demand umwelt- und ressourcenschonend produzi ert.

Bücher schneller online kaufen

www.morebooks.shop

info@omniscriptum.com
www.omniscriptum.com

MIX
Papier aus verantwortungsvollen Quellen
Paper from responsible sources
FSC® C105338

Printed by Books on Demand GmbH, Norderstedt / Germany